BUILDING BLOCKS of COMPUTER SCIENCE

ORDER in CODING

Written by Echo Elise González

Illustrated by Graham Ross

WORLD BOOK

a Scott Fetzer company
Chicago

World Book, Inc.
180 North LaSalle Street
Suite 900
Chicago, Illinois 60601
USA

For information about other World Book publications,
visit our website at www.worldbook.com
or call 1-800-WORLDBK (967-5325).
For information about sales to schools and libraries,
call 1-800-975-3250 (United States),
or 1-800-837-5365 (Canada).

Library of Congress Cataloging-in-Publication Data
for this volume has been applied for.

Building Blocks of Computer Science
ISBN: 978-0-7166-2883-5 (set, hc.)

Order in Coding
ISBN: 978-0-7166-2888-0 (hc.)

Also available as:
ISBN: 978-0-7166-2896-5 (e-book)

Printed in China by RR Donnelley, Guangdong Province
1st printing August 2020

STAFF

Executive Committee
President: Geoff Broderick
Vice President, Finance: Donald D. Keller
Vice President, Marketing: Jean Lin
Vice President, International Sales:
    Maksim Rutenberg
Vice President, Technology: Jason Dole
Director, Editorial: Tom Evans
Director, Human Resources: Bev Ecker

Editorial
Manager, New Content: Jeff De La Rosa
Writer: Echo Elise González
Proofreader: Nathalie Strassheim

Digital
Director, Digital Product Development:
    Erika Meller
Digital Product Manager: Jon Wills

Graphics and Design
Sr. Visual Communications Designer:
    Melanie Bender
Coordinator, Design Development and
    Production: Brenda B. Tropinski
Sr. Web Designer/Digital Media Developer:
    Matt Carrington

Acknowledgments:
Art by Graham Ross/The Bright Agency
Series reviewed by Peter Jang/Actualize
    Coding Bootcamp

TABLE OF CONTENTS

There is a glossary on page 30. Terms defined in the glossary are in type **that looks like this** on their first appearance.

A computer can't run a computer program properly if the instructions are not organized.

That's where I come in handy!

THIS WAY

THAT WAY

THIS WAY

NO! THIS WAY

INSTRUCTIONS

Computer programmers use **control flow** to organize program instructions.

LOOPS CONDITIONS

There are many tools programmers can use to ensure a good control flow.

They can use such programming elements as **loops** and **conditions.**

These elements tell the computer when to carry out program instructions.

Using **loops, conditions,** and other coding elements is an important part of **control flow.**

Programmers use these elements to create different kinds of control flow.

The three basic kinds of control flow are called **sequencing, iteration,** and **selection.**

SEQUENCING
ITERATION
SELECTION

Each of these flows enables the computer to run a program in a particular kind of order.

In a **sequenced control flow,** the computer program runs in the order the code is entered.

The computer simply reads a sequenced program from top to bottom, one line at a time.

Sequenced control flows work well for many straightforward computer programs.

Let's use sequencing to put this burger together.

A sequenced control flow is perfect for this task, because we only need to add the ingredients one at a time until the sandwich is complete.

I'm hungry.

Let's write a program to eat the hamburger we built!

We can use an **iteration control flow** to make this program.

Here are code blocks for taking a bite, chewing, and swallowing.

These steps must be done multiple times, over and over again, until the hamburger is finished.

So, let's use iteration to carry out this part of the code as many times as we want.

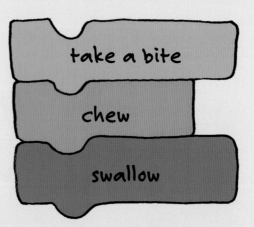

take a bite

chew

swallow

We begin by picking the first baby turtle from the row of turtles. We'll call it Baby Turtle A.

We compare Baby Turtle A to the next turtle in the row, Baby Turtle B.

Baby Turtle A is bluer than Baby Turtle B, so we insert Baby Turtle A into the row TO THE RIGHT of Baby Turtle B.

Then, we compare the next turtle, Baby Turtle C, to the previous turtles in the row. Baby Turtle C is bluer than both baby turtles A and B, so we keep it where it is.

We can continue doing this until all the baby turtles are sorted from the greenest turtle to the bluest turtle!

Another common type of data-sorting algorithm is the **merge sort.**

Merge sorting is a bit more complex than **selection sorting.**

But, it is more useful and efficient in certain situations.

Let's try using a **merge sort** to arrange these parrots.

Polly wants a cracker!

Hi.

This above all: to thine own self be true. And it must follow, as the night the day, thou canst not then be false to any man...

We want the *least* talkative parrots to be on the left and the *most* talkative parrots to be on the right.

Hi.

Polly wants a cracker!

Those friends thou hast, and their adoption tried, Grapple them unto thy soul with hoops of steel,...

With a merge sort, we first divide the group in half.

Then, we split each half again and keep repeating until we can't split the groups in half any further.

So, let's start merging them back together in order from least talkative to most talkative.

Let's compare the first parrots on each branch.

The less talkative one will be the first in the ordered row.

Let's keep comparing the first parrots on each branch...

Until the birds are arranged from shyest to wordiest!

GLOSSARY

algorithm a set of step-by-step instructions used to write computer programs. Algorithms are also used to solve math problems and other problems.

condition a statement that can be true or false. A program may tell a computer to run a piece of code if a certain condition is true.

control flow the order in which a computer follows the steps of a computer program.

data information that a computer processes or stores.

data sorting sorting a list of data into a particular order.

divide-and-conquer a data-sorting strategy in which data is divided, then recombined in a particular order.

insertion sort a sorting algorithm in which the elements in a list are sorted one by one.

iteration a control flow in which certain lines of code are repeated.

loop a piece of code that causes part of a program to run over and over again.

merge sort a sorting algorithm that splits up the elements in a list into groups and pairs, then recombines them in the correct order.

selection a control flow in which certain lines of code are skipped over unless certain conditions are true.

selection sort a sorting algorithm that goes through the elements in a list one by one, choosing the minimum or maximum element.

sequencing a control flow in which the computer program follows the steps of code in order.

sorting algorithm an algorithm that is used to put a list of data in a particular order.

spreadsheet a document in which data is arranged in rows and columns.

GO ONLINE

Now that you know all about how important order is in coding, you can try sorting some data yourself! Go to this website to find the Sort the Giraffes activity, and many other fun computer science activities!

www.worldbook.com /BuildingBlocks

INDEX